WHY DO COWS SLEEP STANDING UP?

ANSWERING KIDS' QUESTIONS

by Nancy Dickmann

PEBBLE
a capstone imprint

Pebble Emerge is published by Pebble, an imprint of Capstone.
1710 Roe Crest Drive
North Mankato, Minnesota 56003

www.capstonepub.com

Library of Congress Cataloging-in-Publication Data
Names: Dickmann, Nancy, author.
Title: Why do cows sleep standing up? : answering kids' questions / by Nancy Dickmann.
Description: North Mankato, MN : Pebble, [2021] | Series: Questions and answers about animals | Includes bibliographical references and index. | Audience: Ages 6-8 | Audience: Grades 2-3 | Summary: "People don't sleep standing up. But cows do! Why? You have questions, and this book has the answers. Find out all about cows and why they don't need to lie down to catch some z's"—Provided by publisher.
Identifiers: LCCN 2020036439 (print) | LCCN 2020036440 (ebook) | ISBN 9781977131690 (hardcover) | ISBN 9781977132765 (paperback) | ISBN 9781977155368 (pdf) | ISBN 9781977156983 (kindle edition)
Subjects: LCSH: Sleep behavior in animals—Juvenile literature. | Cows—Behavior—Juvenile literature.
Classification: LCC QL755.3 .D53 2021 (print) | LCC QL755.3 (ebook) | DDC 599.64/22519—dc23
LC record available at https://lccn.loc.gov/2020036439
LC ebook record available at https://lccn.loc.gov/2020036440

Image Credits
Dreamstime: Nenad Nedomacki, 17; Shutterstock: Apostle, design element, Artem Kniaz, 18, BlueSnap, 6, Burry van den Brink, 11, Clara Bastian, 9, Cultura Motion, 15, Graham D Elliott, 13, Hung Chung Chih, 7, lfdf, 21, owatta, design element, photastic, 20 (index card), prochasson frederic, cover, Raketir, 8, Richard Semik, 19, Sara Winter, 5, Virinaflora, 20

Editorial Credits
Editor: Megan Peterson; Designer: Ted Williams; Media Researcher: Jo Miller; Production Specialist: Spencer Rosio

All internet sites appearing in back matter were available and accurate when this book was sent to press.

Printed and bound in China. 5241

Table of Contents

Words in **bold** are in the glossary.

ASLEEP OR AWAKE?

You visit a farm with your family. Some cows are lying down in a field. Others are walking around as they munch on grass. But then you spot some others. They are standing still. Their eyes are closed. It looks like they're not moving at all. Are they asleep on their feet? How is this possible?

TWO TYPES OF SLEEP

Cows really can sleep standing up! But they're not deep asleep. It's more like a **doze**. Their tails twitch. Their ears move. They will wake up if someone comes near.

There is another type of sleep. It is called **REM sleep**. It is a deep sleep. **Muscles** can't move during this sleep. Humans have REM sleep. Cows also have it. But cows can't do it standing up!

UP OR DOWN?

People need both types of sleep. So do cows! Cows and people must lie down for deep sleep. We can both doze lying down too. But cows can also doze standing up.

Cows lie down a lot. They even do it when they are awake. Cows lying down are often just relaxing. They might also chew their food. Then they burp it up and chew it again! The lump is called **cud**.

SPECIAL LEGS

Could you sleep standing up? You would probably fall over. But cows can do it. This is because of how their legs are made. They can easily stand for long periods.

A cow's knees can "lock" into place. Parts on the inside of the leg hold it straight. This means it doesn't take much **effort** to stand. The cow's muscles don't get tired.

THE STAND-UP SLEEPING CLUB

Imagine you are asleep. Then something scares you! What do you do? Jump to your feet and run away. But cows are heavy. They can't stand up quickly. It is safer for them to doze standing up.

Elephants doze standing up too. So do horses and giraffes. These **mammals** are all large, like cows. They lie down to sleep deeply. But they spend more time dozing on their feet.

I'M NOT TIRED!

You spend about a third of your life asleep. That's a lot of time! Cows need less sleep. They only get about four hours a day. Part of this is REM sleep. They also spend about eight hours dozing. Some of this time they are on their feet.

15

DO COWS DREAM?

Have you ever had a really good dream? Most dreams happen during REM sleep. You might dream about cows. But can a cow dream about you?

Scientists think they probably can. Cows have REM sleep, just like people. But we don't know for sure. A cow can't tell us about its dreams!

QUICK, GRAB AN UMBRELLA!

Can cows tell when it's about to rain? Some people think so! They say that if cows lie down, it will rain soon. This **saying** has been around for a long time.

Scientists don't think it's true. Cows spend a lot of time lying down. It can't always mean rain. So if you want to know if it will rain, watch the news!

SLEEP CLUB TRADING CARDS

What You Need:

- pencil
- paper or card stock
- scissors
- markers

What You Do:

1. Make a list of animals that sleep standing up.
   Some are listed on page 12. Can you research
   more animals that doze on their feet?

2. Research the animals you chose. How tall are
   they? What do they eat? How much do they
   weigh? Find out the same information for
   each animal.

3. Cut paper or card stock into trading cards.
   You can make the cards any size you choose.

4. Draw pictures of the animals on the fronts of the
   cards. Write the facts on the backs of the cards.

5. See if a friend wants to make his or her own
   cards to swap!

Glossary

cud (KUHD)—half-eaten food that an animal burps up and chews again

doze (DOZE)—a type of light sleep; also, to sleep lightly

effort (EH-fuhrt)—the work necessary to do something

mammal (MAM-uhl)—a warm-blooded animal that breathes air and has hair or fur; female mammals feed milk to their young

muscle (MUHSS-uhl)—a part of the body that helps you move, lift, or push

REM sleep (REM SLEEP)—a type of deep sleep during which muscles freeze and most dreams happen

saying (SAY-ing)—a well-known phrase that gives advice or expresses an idea most people believe is true

scientist (SYE-un-tist)—a person who studies the world around us

Read More

Hall, Margaret. *Cows and Their Calves*. North Mankato, MN: Capstone Press, a Capstone imprint, 2018.

Martin, Emmett. *Cows from Head to Tail*. New York: Gareth Stevens Publishing, 2021.

Statts, Leo. *Cows*. Minneapolis: Abdo Zoom, 2017.

Internet Sites

Cool Kid Facts: Cow Facts
coolkidfacts.com/cow-facts/

Farms for City Kids: Cow Fun Facts
farmsforcitykids.org/sbf-jerseys/cow-fun-facts/

Wonderopolis: Do Animals Dream?
wonderopolis.org/wonder/do-animals-dream

Index